Vortical Tectonics: A Theoretical Approach to Tectonic Rotation

J. R. GEMMELL, JR.

March 24, 2019

ABSTRACT

There are various parts of Earth's surface where geologic circles appear. Some of these geologic circles clearly provide evidence of vortical cores of the past and possibly present (e.g., Australia, Fiji, Banda Sea, South China Sea, etc.). There are fractures in the Earth's surface that appear to traverse the globe, which could be members of far greater diameter vortices.

To provide a method of dimensioning for the whole Earth, hurricanes are discussed first. Hurricanes on Earth's surface are considered to be three dimensional. In this case, the Earth's vortical core places two vortices together with one on the top and one on the bottom. The Earth's north pole is the top and the south pole is the bottom. The orientation of the Earth's top and bottom is of no consequence to the manner that we set our paper globes and maps. This study will categorically identify clearly evident lithospheric vortices and suspect vortices. To help identify these vortices, fractures and fault lines emanating from their centers will be mapped.

Subject headings: vortical tectonics

INTRODUCTION

To understand the complex nature of vortices, for the purpose of this study, all bodies in space shaped like a ball will be given a primary categorical name. This primary categorical name is *Rotating Celestial Body* (RCB). Not all RCBs have rotations, but were developed into spheres by rotations. Some RCBs are tidally locked in orbits around their parent RCB.

The RCB being addressed in this document is Earth. Earth, however, takes the RCB concept one step farther. This extra step in categorizing RCBs is to identify which RCBs possess differential rotation. For the purpose of this study, RCBs possessing differential rotation will be called *Differentially Rotating Celestial Bodies* (DRCBs). Stars and gas giants are permanent examples of DRCBs. Terrestrial planets might have begun as DRCBs, but if they become tidally locked with a parent RCB, like our Moon being tidally locked into orbit around the Earth, then they no longer hold the status of RCB, but may still be differentially operable. A Non-RCB's ability to remain differentially operable, yet not be rotating may be a source for a dynamo.

First, the Earth will be broken down into six vortices. These six vortices are the only vortices possessing vortical rods that connect to an anti-vortex on the other side of the planet.

The first vortical set of two is the Earth's polar axis. The Earth's polar axis is the primary axis. The second axis is a vortical set running along the 30° longitude extending from east to west at the equator. The third set runs along the 120° longitude axis from east to west at the equator. This might not be a mere coincidence of geometry, but a DRCB geophysics principle of operation.

A FOURTH-DIMENSIONAL PERSPECTIVE OF EARTH

When we think of hurricanes on the Earth's surface, we consider them to be three dimensional in nature. A hurricane's three dimensions may be referred to as the *length, height and depth*. A hurricane's greater vortical diameter exists at the top of its three-dimensional cube. Its smaller vortical diameter is located at its base, or at the surface of the Earth. Gravitational force is stronger on the Earth's surface, closer to its core, than at higher altitudes in the Earth's atmosphere. The Earth generates its gravitational force from the center of its vortical core. Gravity is stronger at the vortical core and weakens as it emanates from the vortical core into space.

The Earth's vortical core extends from the north pole down to the south pole. The Earth's core rotates with the polar vortical core in a prograde direction. If we separate the northern hemisphere from the southern hemisphere, we will see opposites. Cut a globe in half and set each half side by side on a table. Arrange the setting so that the northern hemisphere is before the observer and to the left of the southern hemisphere. Knowing the Earth's rotation, we can see the southern hemisphere rotates in a retrograde direction while the northern hemisphere rotates in a prograde direction. However, mirrored to each other, both poles rotate in a single prograde direction.

Space has no ground to establish a base. Space is omni-directional applying pressure spherically around a single coordinate that forms into the shape of a RCB. Therefore, the core of a RCB becomes the base, fusing two vortices into a single *fourth-dimension vortex*. The core of this fourth-dimensional vortex becomes a *vortical rod*.

CHOOSING THE RIGHT VORTEX

During this study, the locations of over 200 vortices have been plotted. This number of vortices includes suspect vortices as well as the most evident vortices. To begin this study, the two most evident vortices are examined. As mentioned before, vortices are extraordinarily complex. Some vortices rotate in opposite directions. Some vortices exist within the confines of other, larger vortices. Vortices also appear to have the quality of existing in binary relationships. A vortex existing inside a larger vortex may be evidence of sequential vortical activity or concurrent vortical activity.

On Earth, opposite vortices are seen with hurricanes in our atmosphere. We have categorized these hurricanes into two phenomena. The two hurricane categories of rotation may be referred to as cyclonic and anticyclonic. Cyclones rotate in a counter clockwise direction in the northern hemisphere. Anticyclones rotate in a clockwise direction in the northern hemisphere. The opposite occurs in the southern hemisphere when the observer views the planet like a globe set on the top of a desk. If we place a globe upside down on the top of a desk, the hurricane rotations in the southern hemisphere would appear to operate the same as the northern hemisphere (e.g., Coriolis Effect).

Lithospheric and polar vortices will most generally present two opposite Spiral Density Faults (SDFs). Some vortices appear to have a primary or single SDF. Fiji is a prime example of a vortical core clearly possessing two SDFs (see fig. 1). As a general rule, when analyzing planetary terrain features, faults can be traced in terms of identifying terrain features lying in a linear formation. Linear formations most generally, but not always, follow a spiral density path that runs on an expanding tangent from the vortical core. In some instances, faults can be transformed along fractures fragmenting their relief. In other instances, faults can be curved by bending in competing lithospheric pressure zones. Vortices do not appear to be the only factor in tectonics. A single SDF can possess several different forms of terrain features. Often, a broad view of a fault line is necessary to identify complete paths. In this study, fault line features are: fractures (e.g., transform and divergent), troughs, escarpments, rifts, seamounts, subterranean ridges, etc. Faults are further categorized as: primary, intermediate or inferior. Sedimentary partially concealed rifts are an example of inferior fault lines. Partially concealed seamounts and/or subterranean ridges comprise features that are considered to be intermediate. Fractures are primary.

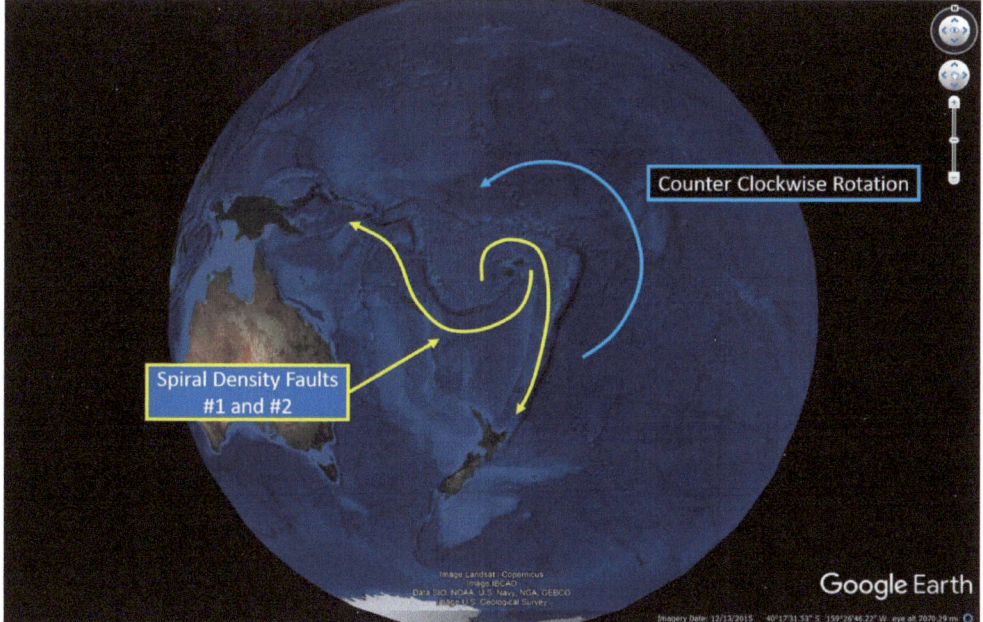

Fig. 1. — This figure outlines Fiji's two prominent spiral density faults and presents the island's rotation as counter clockwise.

The two most prominent vortices visible on the Earth's surface are Lake Victoria in Africa and the North Banda Basin in Indonesia (see fig. 2). The intent of this study is to use these two vortices and their anti-vortical companions to establish a control. In the process of the study, it was discovered that these two special vortices have far reaching rotational perimeters. They are at approximately 90° intervals apart from each other and have vortical pairs on the opposite side of the planet. Hence, these two vortices *may* be all that is needed to establish a control for this initial experimental study.

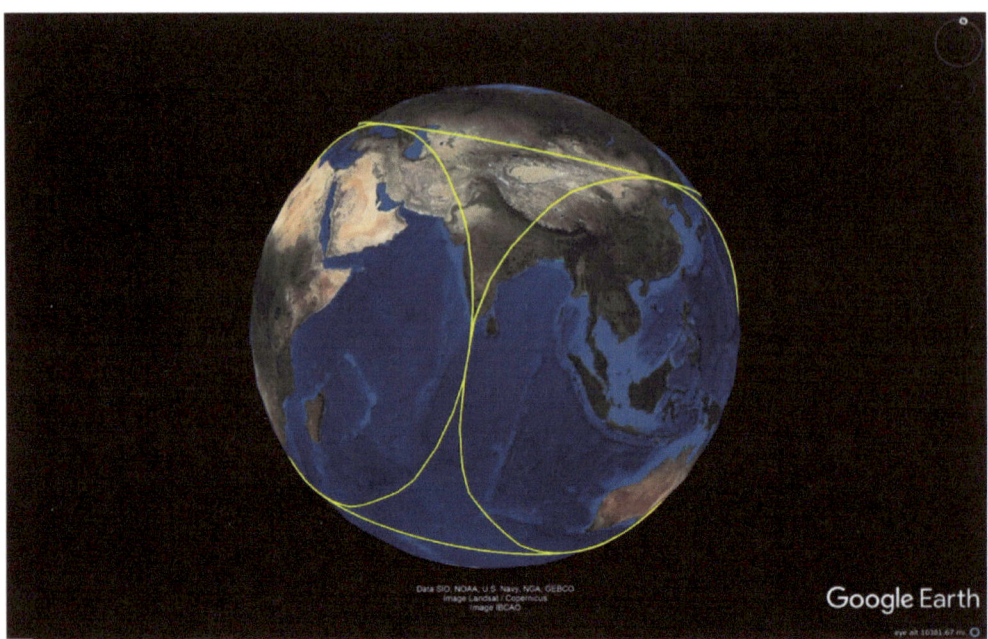

Fig. 2. — This figure presents four of the 45° vortical perimeters that includes the 30° and 120° equatorial axes, and the Arctic and Antarctic axis. An element that raises curiosity, is what appears to be a general alignment of the west and east shorelines of India relative to the tangent of the African and Indonesian vortical perimeters.

THE FOUR EQUATORIAL ROTATIONS

Three of these equatorial vortices appear to have depressions at their vortical cores. The Pacific Ocean is much more elusive, possibly due to oceanic sedimentary concealment. However, the Pacific Ocean vortex does not appear to have a relief geologic feature either. Hence, for the purpose of this study, a hypothesis of the Pacific Ocean's vortical core possessing a depression is assumed. Lake Victoria, in Africa is examined first. Then, the North Banda Basin, Pacific Ocean, and Sipaliwini are examined in the order listed in this paragraph.

The Lake Victoria Vortex

Lake Victoria is the eastern hemisphere vortical core of the 30° vortical rod extending from east to west parallel to the equator. The Lake Victoria Vortex (LV2) is rotating in a clockwise direction in the southern hemisphere. The LV2 is cyclonic in terms of a rotational direction, but possesses a depression at its center which may be implicative of an anticyclone's operation in principal. If LV2 is anticyclonic in principal, it may be possible the vortex has a higher degree of pressure gradient above the lithosphere and a lower pressure gradient below. The LV2's coordinates are: 0°46'56.87" South Latitude, 32° 6'57.70" East Longitude (see fig. 3). The LV2 is located below the equator. As with any vortex, a vortical core might not always be central to a vortex, but migrating in a vortical precession.

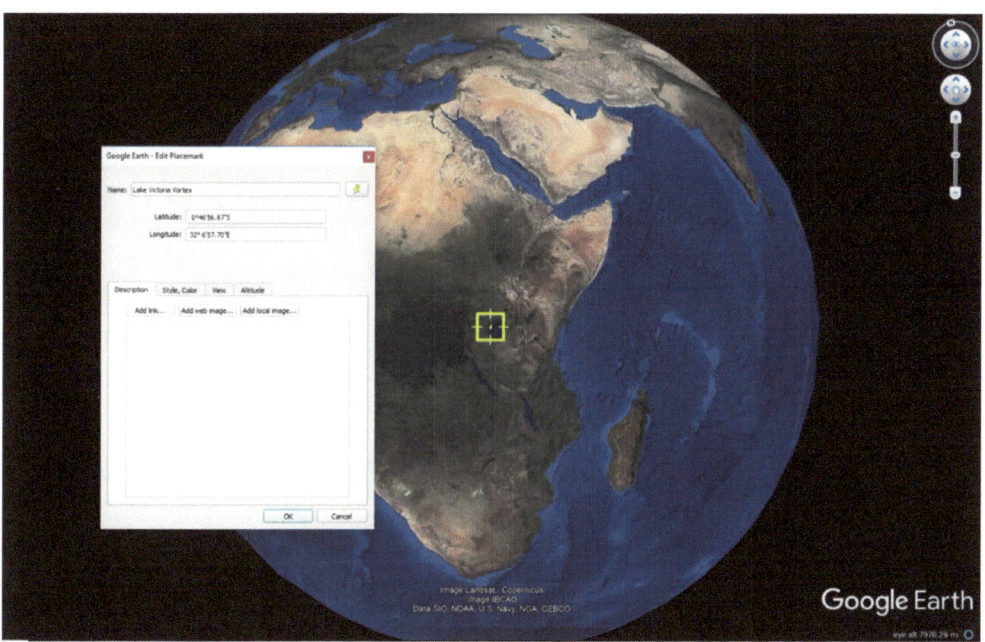

Fig. 3. — This figure presents the "Edit Placemark" dialogue box displaying the coordinate to the LV2's vortex as plotted by intersecting the center of the 45° vortical radius. Vortical cores may not always be found at Top Dead Center (TDC) of a vortical diameter, and most likely migrate in a lithospheric form of precession.

The LV2 possesses a single, strongly evident SDF (see fig. 4a). The LV2 may be splitting into two or even three SDFs (see fig. 4b). The LV2 intermediate SDF might be plotted as shown in figure 4c below. The LV2's 45° radius perimeter is shown in figure 4d. The north eastern tangent of LV2's perimeter aligns with India's lower western coast along the Arabian Sea. The perimeter's tangent proceeds northward bordering the northern arc of the Caucasus Mountains in Russia. It passes along the northern escarpment of the Black Sea, separating the Black Sea from the Sea of Azov.

Africa, consuming most of the eastern 30° vortex may have spun in a retrograde direction from -158° to -199°, a difference of -41°. This is an estimate of continental migration following the breaking down of Pangaea some 4.4 billion years before present. General age of Earth is established based on an article written by Oskin (February 23, 2014), titled "Confirmed: Oldest Fragment of Early Earth."

Fig. 4a. — This figure presents LV2's most prominent SDF. There is a possibility that LV2's SDF #1, shown in this figure, splits into two SDFs.

Fig. 4b. — This figure presents a possible deflection in LV2's primary SDF shown in the color of fluorescent blue.

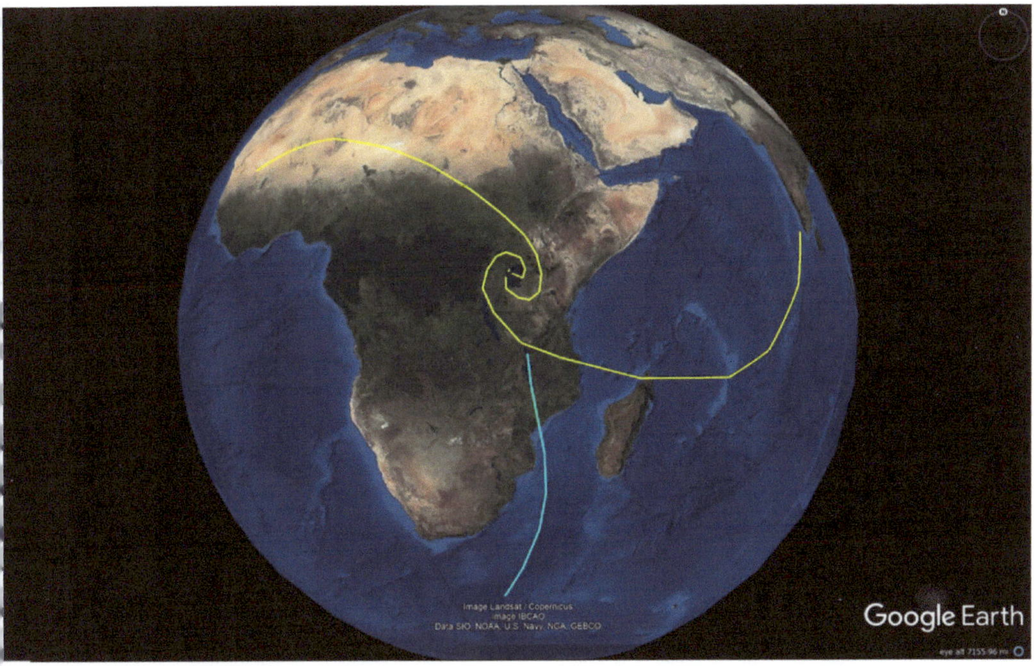

Fig. 4c. — This figure presents three of LV2's SDFs.

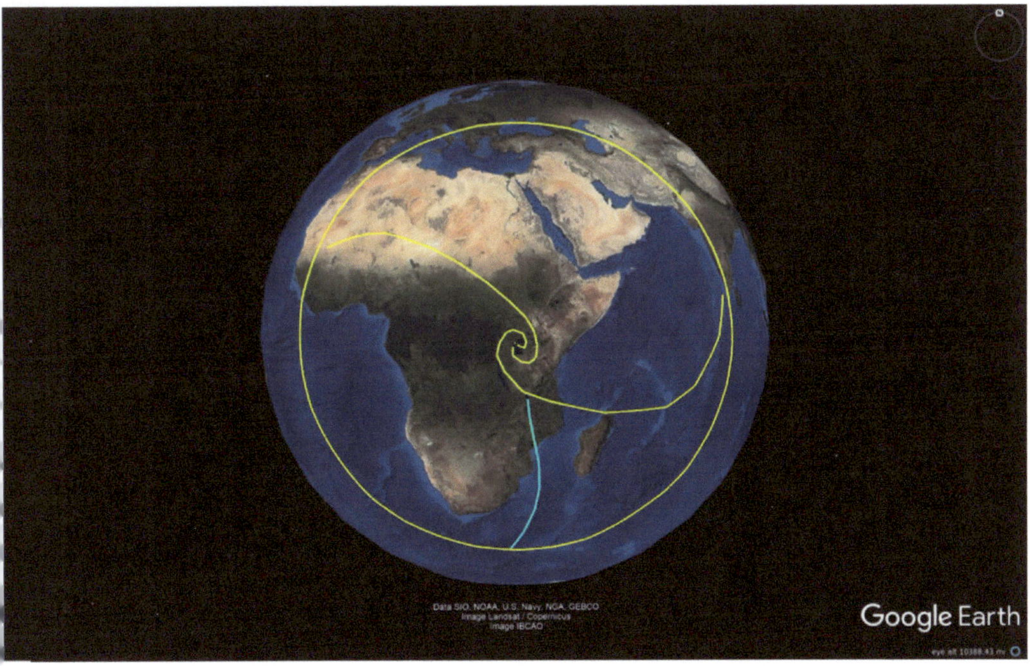

Fig. 4d. — This figure presents LV2's 45° vortical perimeter and its SDFs.

The North Banda Basin Vortex

The North Banda Basin is the eastern hemisphere vortical core of the 120° vortical rod extending from east to west parallel to the equator. The North Banda Basin Vortex (NB2V) is also rotating in a clockwise direction in the southern hemisphere. The NB2V is cyclonic in terms of a rotational direction, but possesses a depression at its center which may be implicative of an anticyclone's operation in principal. The NB2V's coordinates are: 3°50'10.44" South Latitude, 124°54'42.95" East Longitude (see fig. 5). The NB2V is also located beneath the equator. In determining the NB2V's rotation, in principal, it must rotate in an opposite direction of its antivortex. South America is challenging to identify its vortical core, but the Blue Ridge Mountains in North America form terrain along the Sipaliwini Vortical perimeter supporting the rotational analysis of the Sipaliwini Vortex. The NB2V center was identified by intersecting the center of the eastern hemisphere's 120° vortical perimeter. An exceptional analysis also presents the NB2V, not only as a depression, but a depression in the shape of a pentagon. This pentagonal configuration is similar to Jupiter's south pole vortex, however, identifying the five sub-vortices is difficult under the current geological circumstances involving the unique features established in the Indonesian vortex.

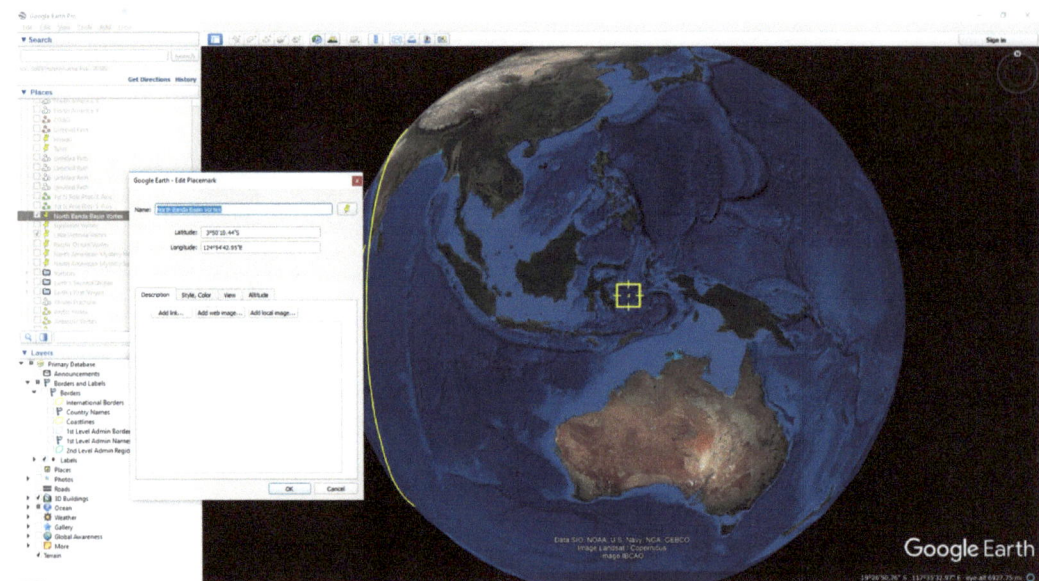

Fig. 5. — This figure presents the "Edit Placemark" dialogue box displaying the coordinate to the NB2V's vortex as plotted by intersecting the center of the 45° vortical radius.

Just east and south below the NB2V lies another more prominent vortex (see fig. 6). This is the vortex that rotates in the same direction as the Sipaliwini Vortex in South America. At this time, it is hypothesized that other forces have influenced the geological deformation of this apparent binary vortical relationship existing in Indonesia. A binary vortical relationship whereby each vortex is rotating counter to one another. The NB2V's binary companion's SDF's are much clearer in presence and may have occurred more recently than the NB2V. This recent activity may have compromised or eliminated the visual presence of the five sub-vortices expected to be found surrounding the NB2V's pentagon.

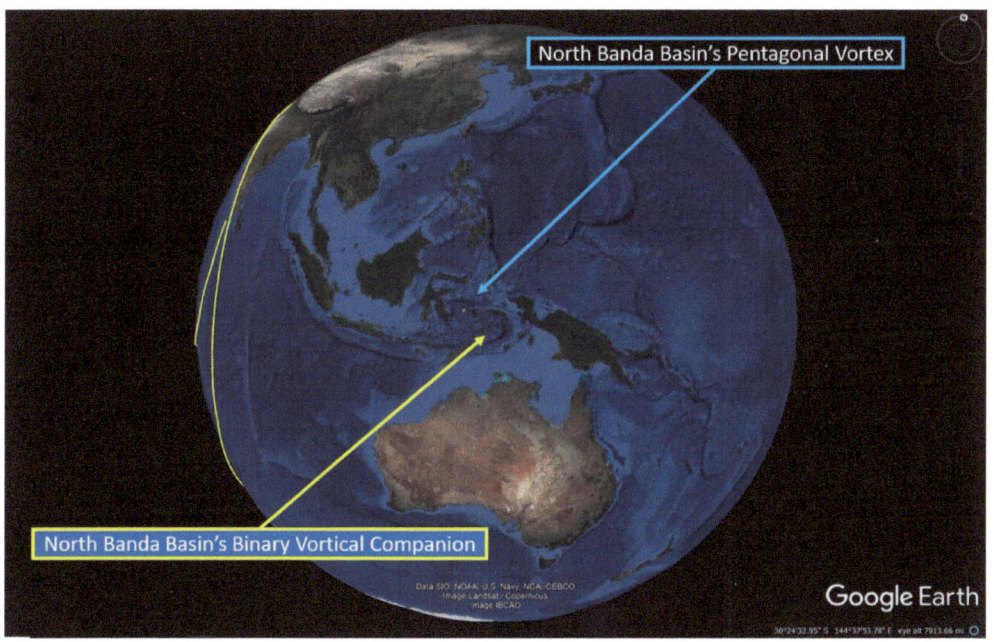

Fig. 6. — This figure presents the NB2V and its binary companion to the southeast.

The NB2V's SDFs are presented in figure 7 below. Figure 8 below displays NB2V's 45° vortical perimeter. The vortical perimeter of NB2V appears to run in tandem with the eastern shoreline of India, before merging on a tangent with the eastern hemisphere's 30° vortical perimeter. The apparent mergence of the two vortices may suggest a geological explanation for India's continental formation. The NB2V's perimeter passes through the Tsugaru Strait separating Hokkaido, Japan, from Honshu, Japan. Then, the perimeter bisects the Himalayas. The southwestern perimeter passes through the Bass Strait separating Tasmania, Australia, from Victoria, Australia.

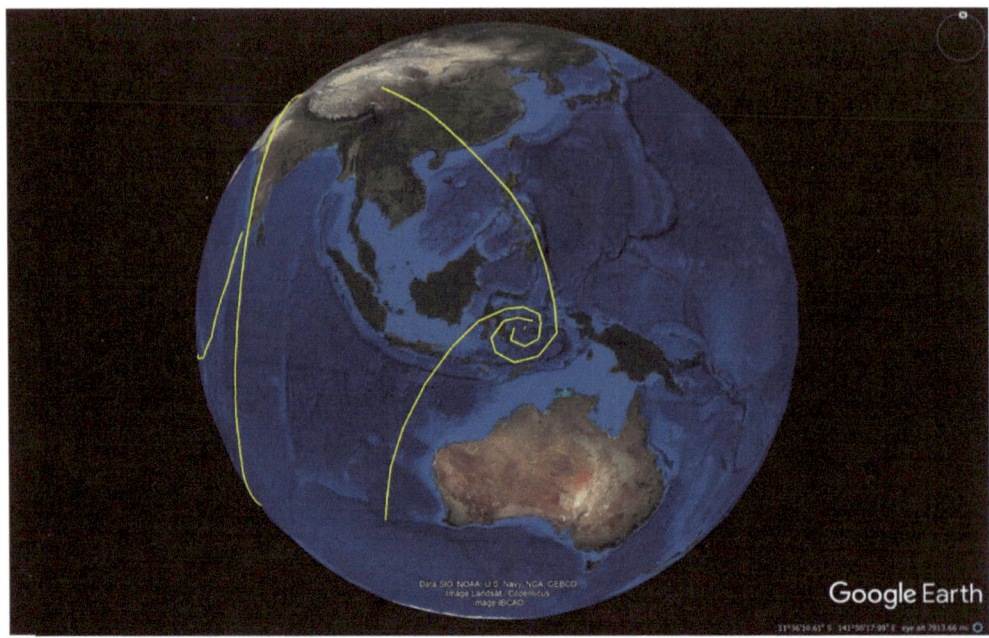

Fig. 7. — This figure presents NB2V's SDF #s 1 and 2.

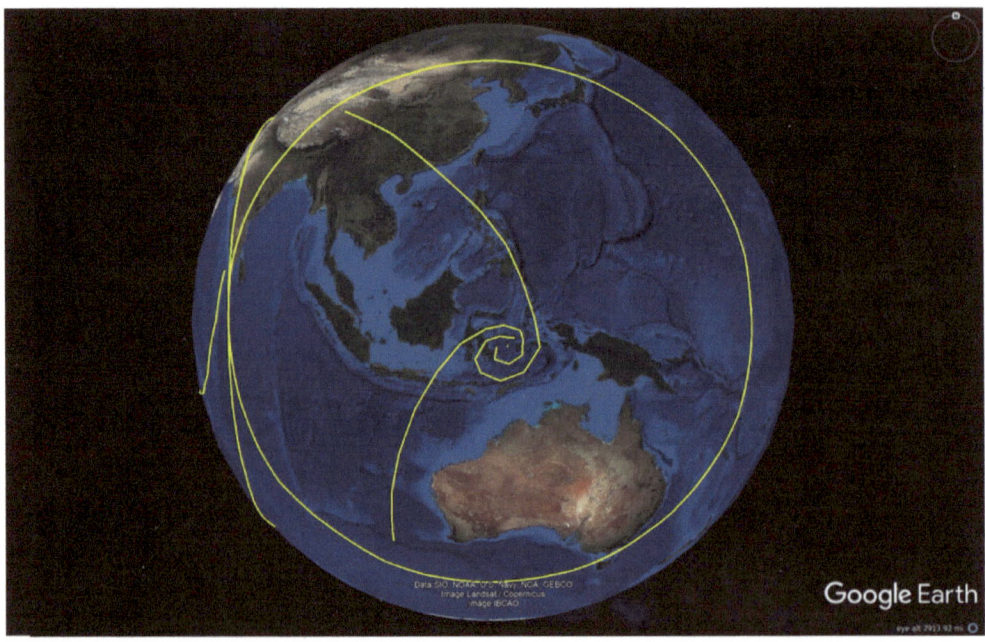

Fig. 8. — This figure presents NB2V's 45° vortical perimeter and its SDFs. The NB2V's northwest perimeter tangent appears to run in tandem with the eastern shoreline of India along the Bay of Bengal.

The Pacific Ocean Vortex

The Pacific Ocean consumes almost the entire western hemisphere's vortical core of the 30° vortical rod extending from west to east parallel to the equator. The Pacific Ocean Vortex (POV) is rotating in a counter clockwise direction in the northern hemisphere. The POV is cyclonic in terms of a rotational direction, but likely possesses a depression at its center which may be implicative of an anticyclone's operation in principal. The POV's coordinates are: 0°46'56.87" North Latitude, 147°53'2.31" West Longitude (see fig. 9). The POV is located above the equator. The POV center was identified by intersecting the center of the western hemisphere's 30° vortical perimeter.

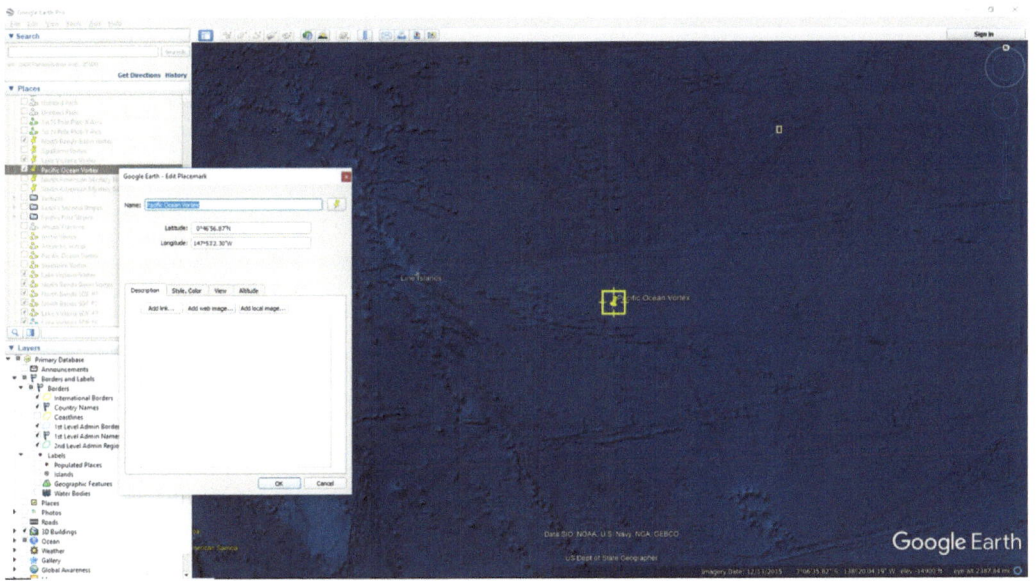

Fig. 9. — This figure presents the "Edit Placemark" dialogue box displaying the coordinate to the POV's vortex as plotted by intersecting the center of the 45° vortical radius.

The POV's SDF #s 1-3 are presented in figure 10 below. In this instance, due to extremely inferior fault lines in the Pacific Ocean, the San Andreas fault is used to establish a control in determining the POV's rotation. This supposed rotation is complimentary to the LV2 rotation. Figure 10 displays an analysis of the POV's SDFs based on available terrain features. Circumstantial evidence suggests the vortex's rotational direction. The 45° perimeter radius of the POV cuts through the middle of the Gulf of California following along the San Andreas Fault and returning to the Pacific Ocean just north of San Francisco, California (see fig. 11). Based on our understanding of the direction of movement along the San Andreas Fault, it may be presumed that the POV is rotating in a counter clockwise direction. This counter clockwise directional rotation moves northward on the western side of the San Andreas Fault, while the North American continent is rotating in a southward direction. The San Andreas Fault also lies within the Sipaliwini Vortex which is rotating in the same direction as the POV.

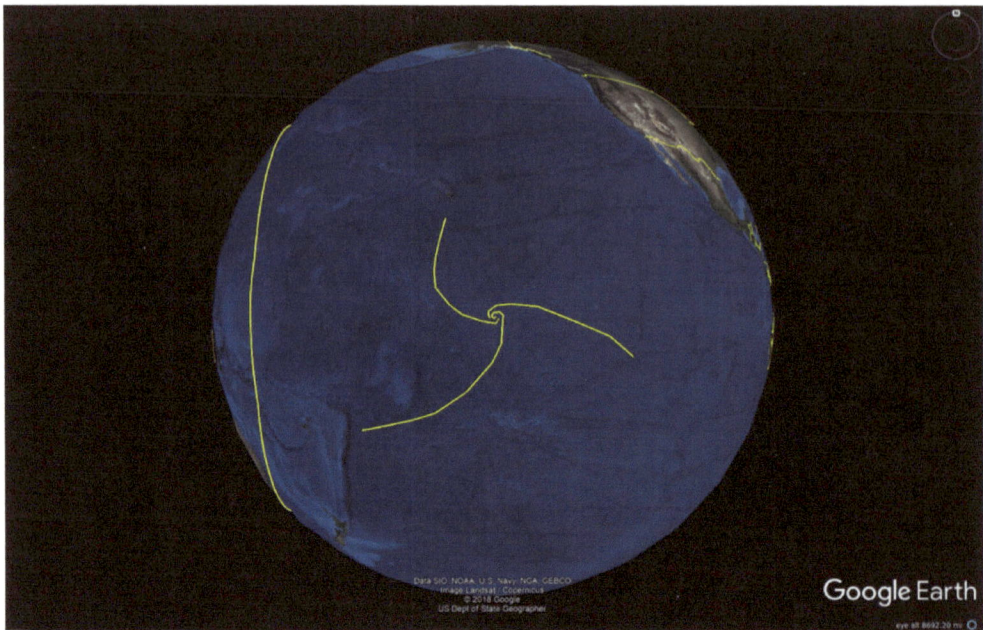

Fig. 10. — This figure presents POV's SDF #s 1-3.

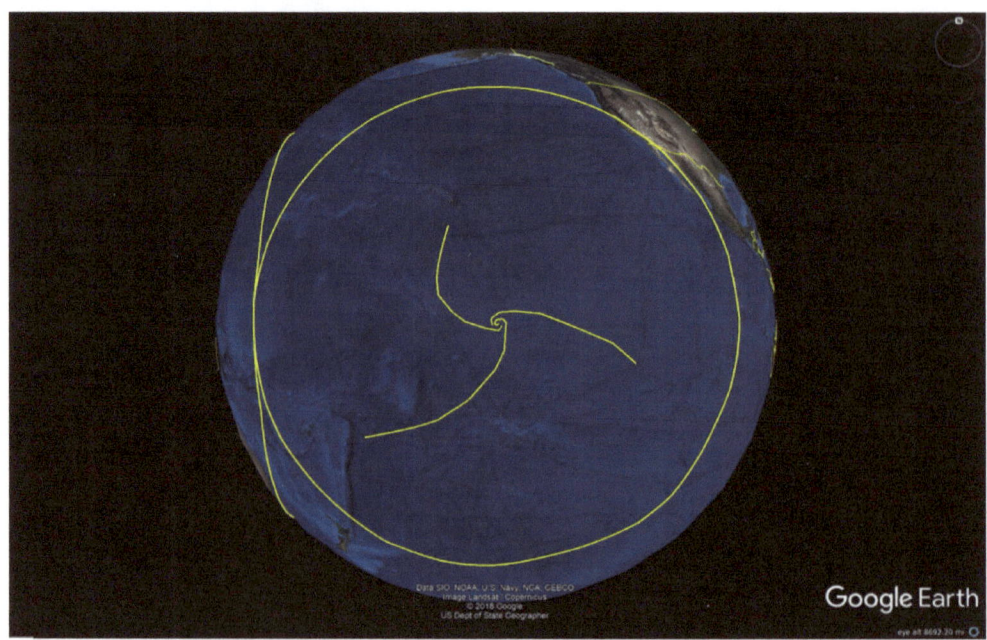

Fig. 11. — This figure presents the POV's 45° vortical perimeter and its SDFs. The POV's northeast perimeter tangent appears to run in tandem with the San Andreas Fault and angular formation of the western United States of America. The southwest perimeter appears to align with the northwestern SDF of Fiji Island.

The Sipaliwini Vortex

The Sipaliwini Vortex (SV) is located above the equator in the western hemisphere. It is the western hemisphere's 120° vortex. Though, much more difficult to locate and identify, the SV's SDFs *can* be seen. Like the San Andreas Fault assisting in determining information regarding the POV, the Blue Ridge Mountains in the eastern United States of America assist in confirming South

America's vortical rotation. The SV coordinates are: 3°50'10.43" North Latitude, 55° 5'17.10" West Longitude (see fig. 12).

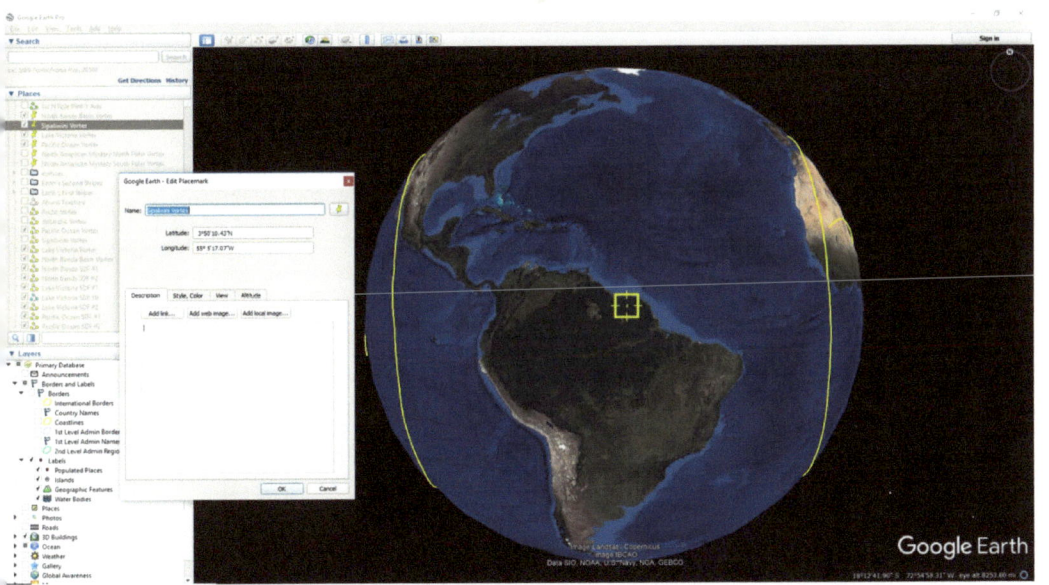

Fig. 12. — This figure presents the "Edit Placemark" dialogue box displaying the coordinate to the SV's vortex as plotted by intersecting the center of the 45° vortical radius.

The SV SDFs are presented in figure 13 below. As mentioned earlier, sometimes taking a broader perspective of the Earth unveils its hidden secrets. The North American and Gröll possibilities present such a perspective (see fig. 14). In figure 15, the vortical perimeter of the SV is presented. The continental shaping of the North American continent is complimentary to the vortical rotation of the SV.

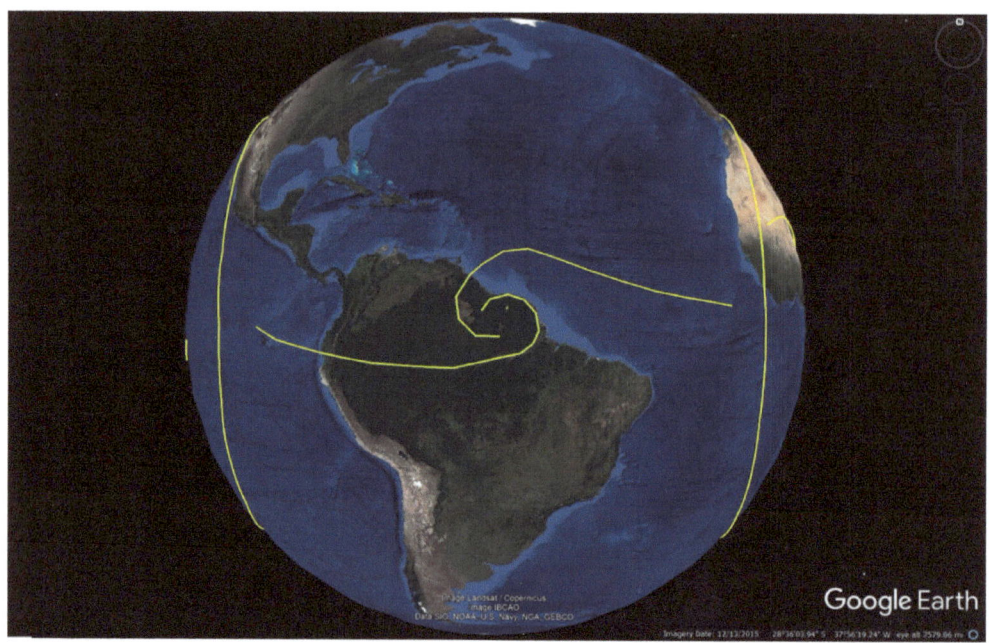

Fig. 13. — This figure presents SV's SDF #s 1 and 2. Further analysis may even suggest the possibility of another SDF wrapping around the Caribbean and running along the eastern shoreline of the United States of America. There is also evidence of a counter SDF opposite the possible North American SDF that streams out along an oceanic rift that cuts through the Gröll Seamount extending southeast off of the Pernambuco Plateau in South America.

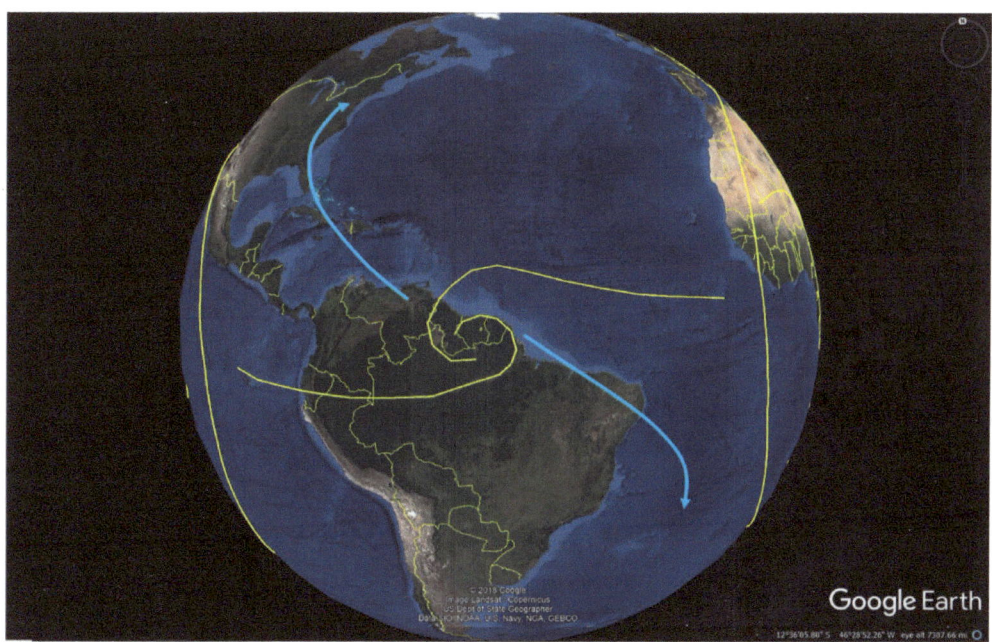

Fig. 14. — This figure presents SV's possible SDF #s 3 and 4 identified as fluorescent blue.

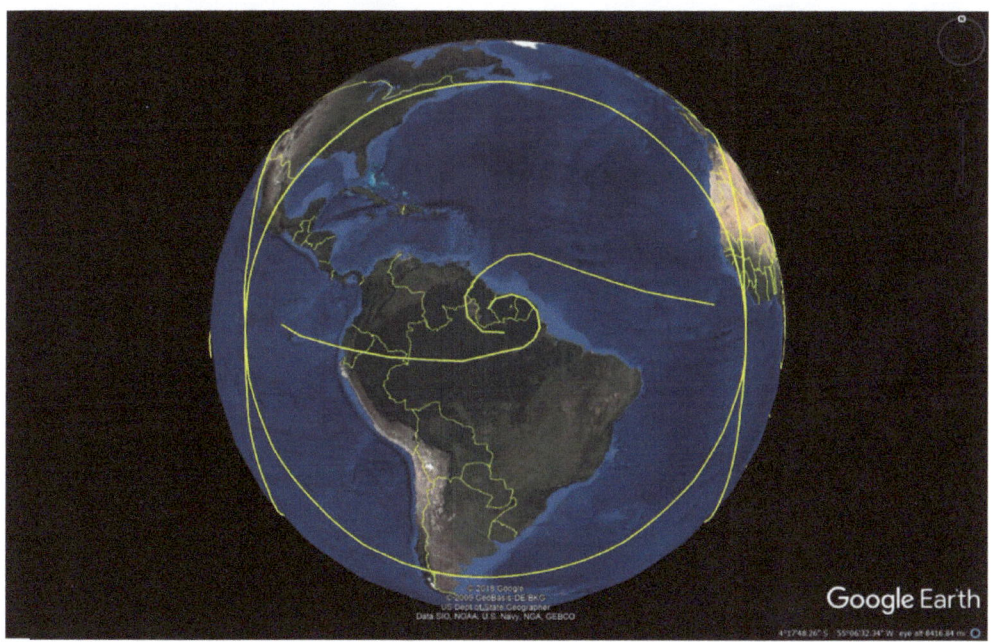

Fig. 15. — This figure presents the SV's 45° vortical perimeter and its SDFs. This vortical perimeter can be seen following the western border of the Appalachian Mountains, then somewhat aligning with Mexico's continental formation before merging into its tangent with the POV.

THE ANTARCTIC VORTEX

The Antarctic Vortex (AnV) is rotating in a counter clockwise direction (see fig. 16). Figure 17 displays the AnV's 45° vortical perimeter.

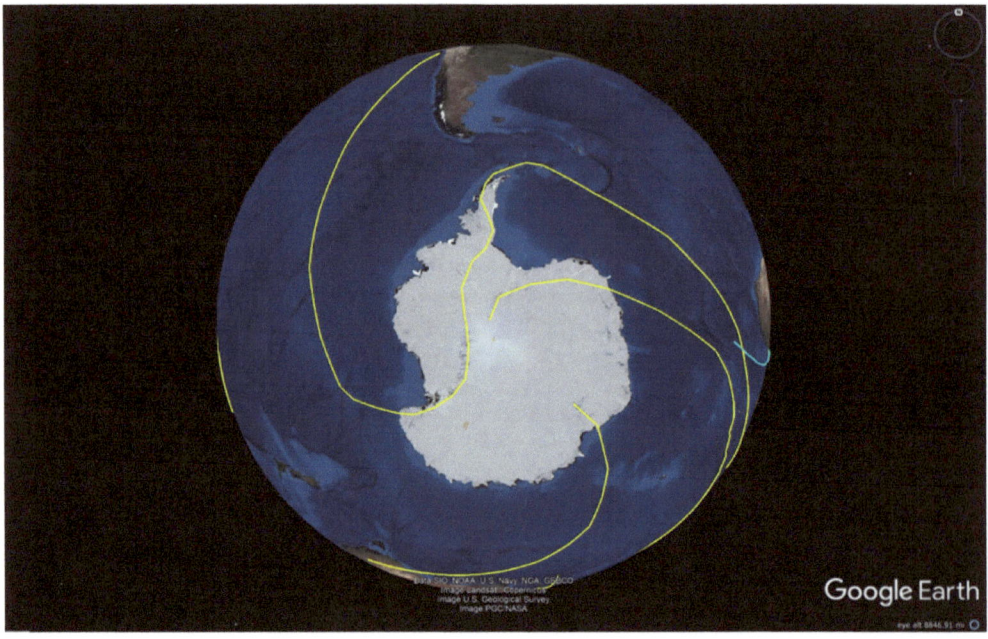

Fig. 16. — This figure presents SV's SDF #s 1-4. Antarctica's SDFs are the farthest reaching. Not shown here, one fault line can be traced extending from the Antarctic Circle straight to the San Andreas Fault in southern California. Whether or not this fracture is indeed related to the San Andreas Fault, is difficult to say, and it will take instrumentation and core samples to prove its correlation. Using the AnV, we can surmise the North Pole is also rotating in the opposite direction of the Earth's rotation.

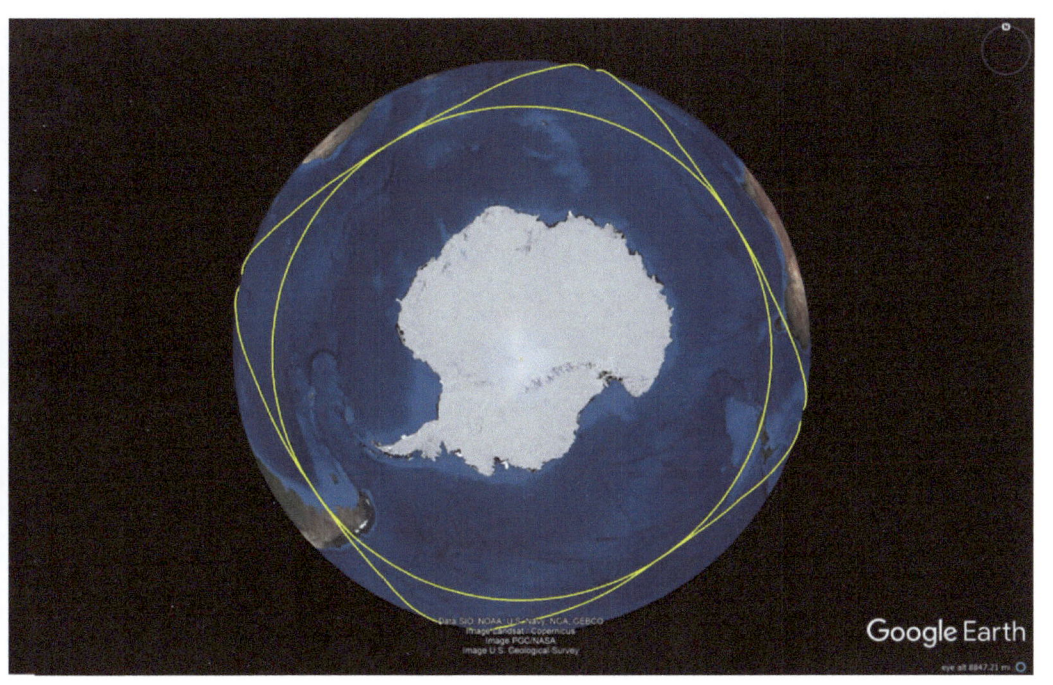

Fig. 17. — This figure presents the AnV's 45° vortical perimeter without its SDFs.

THE ARCTIC VORTEX

Figure 18 displays the Arctic Vortex's (AnV's) 45° vortical perimeter.

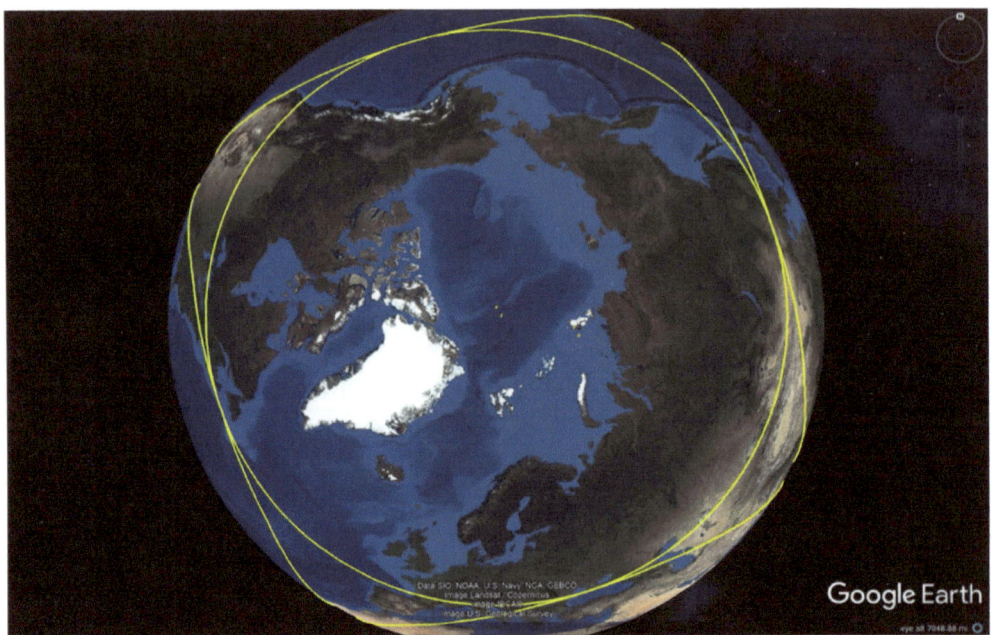

Fig. 18. — This figure presents the ArV's 45° vortical perimeter without its SDFs.

DISCUSSION

A paradox is discovered in determining whether or not a lithospheric vortex is cyclonic or anticyclonic. Lithospheric vortices appear to have the opposite theory of operation than atmospheric vortices. Lithospheric vortices possessing depressions at their vortical cores suggest they are anticyclonic. Lithospheric vortices possessing reliefs at their vortical cores suggest *they* are cyclonic. Antarctica and Fiji are examples of cyclonic lithospheric vortices. Lake Victoria and the North Banda Basin are examples of anticyclonic lithospheric vortices.

Antarctica is visually rotating in the opposite direction as the Earth's rotation. Since the Earth is rotating in a prograde direction, because it is rotating in the same direction as the Sun, then we can postulate Antarctica is rotating in a retrograde direction. When Antarctica is at the center of our planetary view, it appears to be rotating in a counter clockwise direction. Since Antarctica's rotation is counter to the Earth's rotation, then we might assume the Earth is rotating in a clockwise direction.

If we view the Earth's north pole, the Earth will appear to be rotating in a counter clockwise direction, same as Antarctica. How can this be if the two rotations counter each other from an equatorial perspective? The control that must be established is whether or not a vortex is rotating in a clockwise or counter clockwise direction as seen from the observer's TDC planetary view.

The Earth's polar axis is a fourth-dimension vortex whereby the planetary body is rotating in a prograde direction when seen rotating on its axis along its orbital path. From the same perspective, the north and south pole vortices are rotating in the opposite direction, contrary to the Sun's rotation, and hence, are retrograde. If the observer views the Earth's north pole from a position that is TDC, the Earth will be rotating in a counter clockwise direction while the ArV is rotating in a clockwise direction.

Theory of an Inverse Differential

The Earth's polar axis is comprised of the ArV and AnV. These vortices are the primary drivers of *vortical torsion* lying between the Earth's polar axis and the two equatorial axes. A secondary force acts on the type of vortex (e.g., cyclonic/anticyclonic). The larger a vortex is, the more its operational properties are influenced by the secondary force. A smaller vortex's cyclonic activity is influenced by a tertiary force. The tertiary force is not strong enough to influence hemispheric vortices, and the secondary force is unable to influence the smaller vortices.

The LV2 and NB2V reside in the southern hemisphere in closer proximity to the AnV. The AnV's rotation is counter to the LV2 and NB2V rotations acting as a gear. The POV and SV reside in the northern hemisphere in closer proximity to the ArV. Likewise, the ArV acts as a gear driving these two vortices to rotate in an opposite direction. This particular theory suggests the vortical torsion is centric to the Earth's core and therefore differential in operation. The differential vortical operation is the mechanism that determines the rotational direction of a hemispheric vortex.

In addition to the ArV's and AnV's differential influence of vortical rotation to their nearest proximity companions, is the theory of the second force. The second force is above the Earth's lithosphere and crust. The Earth's atmosphere is strong in terms of jet streams and wind, but hardly strong enough to manipulate and mold the lithosphere into spiral vortices. The second force is solar wind pressure. Solar wind pressure is one of three forces that influence the direction of planetary rotations in our solar system. This same force influences the anticyclonic activity of hemispheric vortices. This particular theory of operation is opposite of differential, and can be referred to as an inverse differential. Vortical differential operation comes from within the Earth and vortical inverse differential operation comes from the Sun.

Earth is Gyroscopic

When the observer views the Earth with a north/south orientation perpendicular to the equator, from a perspective that is TDC over the LV2, the rotations of the NB2V and SV are one in the same with the wave action roll rolling away from the observer (see fig. 19). When the observer views the Earth with a north/south orientation perpendicular to the equator, from a perspective that is TDC over the NB2V, the rotations of the LV2 and POV are one in the same with the wave action roll rolling toward the observer (see fig. 20). In conclusion of the equatorial vortices' direction of rotations relative to each other, they are opposite.

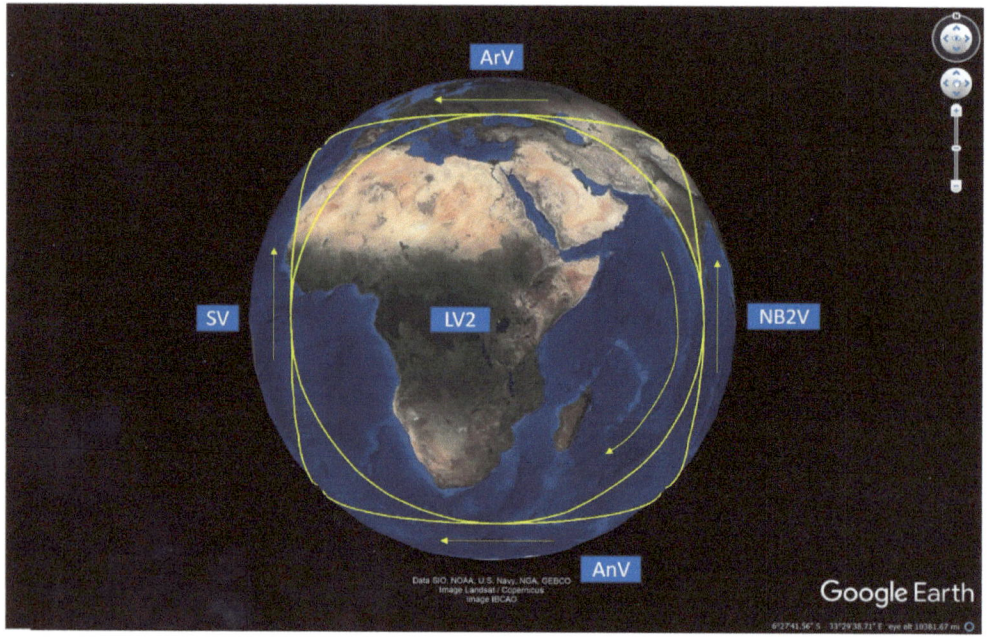

Fig. 19. — This figure presents the axial rotation of the 120° vortical rod in relation to the vortical rotation of the LV2.

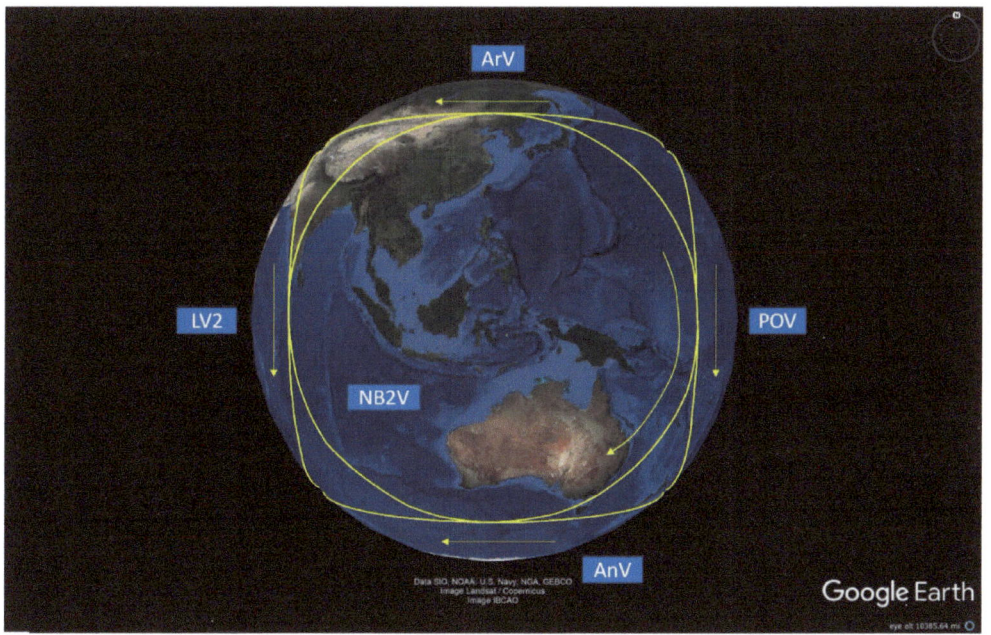

Fig. 20. — This figure presents the axial rotation of the 30° vortical rod in relation to the vortical rotation of the NB2V.

The Earth's polar axis, controlled by the polar vortices, generates the Earth's primary gravitational field. The Earth's primary gravitational field irradiates from the Earth's core out in a wave action design along the Earth's rotational axis. The 30° vortical rod and 120° vortical rod rotations also generate gravitational forces. The equatorial vortical rods cancel each other out as their gravitational outputs clash. Though the equatorial vortical rods cancel-out each other's gravitational influence, they still function as balancing rods holding the Earth soundly centered on its polar axis.

SUMMARY

According to Lee (December 7, 2013), the 2011 earthquake near Japan occurred in the Japan Trench. Tsuji, et al (2019) presented a graphic slide in his report depicting the location and direction of fault slippage in the Japan Trench (see fig. 21). The Tsuji, et al. (2019) report supports the vortical tectonic theory of operation, as well as the gyroscopic planetary motion of hemispheric vortices in conjunction with the theory of vortical torsion.

There is yet much more to learn about how the Earth operates. This study continues into the next phase of planetary terrain analysis that identifies the functionality of *litho streams* and their impact on vortical rotation in the lithosphere. Lithospheric vortices are examined to establish an understanding of their electromagnetic impact on the Earth's agricultural and atmospheric environments. The mechanisms that influence vortical tectonics and the forces that operate behind continental migration over the surface of the lithosphere will be investigated.

Ultimately, a study to explain a theoretical planetary state of worldwide liquefaction will be conducted. Worldwide liquefaction may be responsible for theoretical asthenospheric tsunamis that travel beneath the continents and lithosphere until emerging through lithospheric floors and crustal plates in the form of semi-molten waves. The Solar Dynamics Observatory (SDO) video recorded global tsunamis in the Sun's corona in 2010 giving rise to the possibility of such events taking place on Earth.

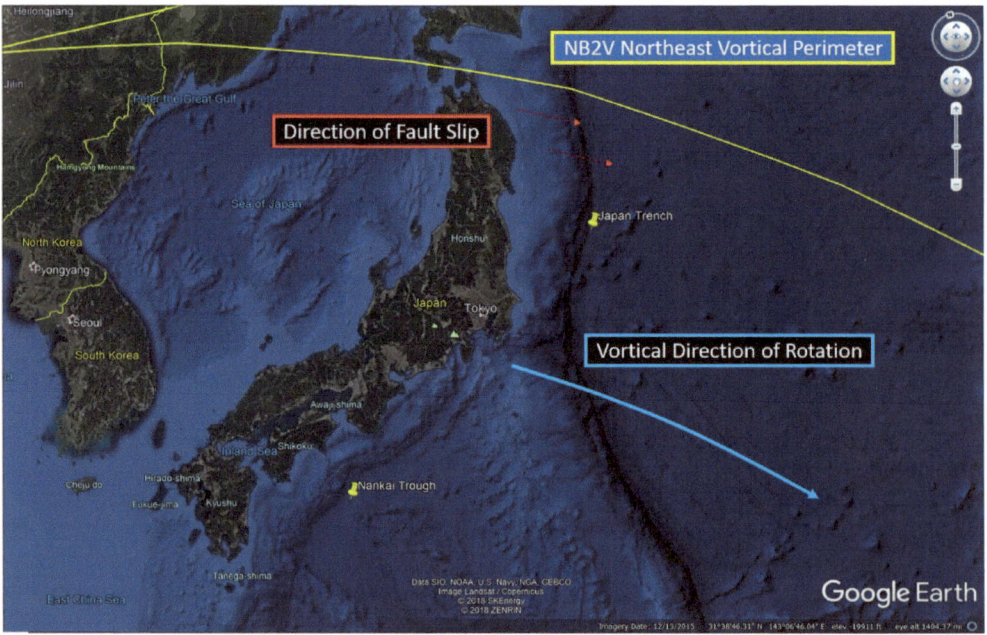

Fig. 21. — This figure presents the NB2V's vortical rotation in relation to the Japan Trench and Nankai Trough. The red arrows denote the direction of fault slippage according to Tsuji, et al. (2019).

INDEX

Antarctic vortex	16
Arctic vortex	18
Asthenospheric tsunami	21
Differentially rotating celestial body	1
Discussion	18
Fourth-dimensional vortex	2
Gyroscopic	19
Inferior fault	3
Intermediate fault	3
Inverse differential	19
Japan Trench	21
Lake Victoria vortex	3
Litho stream	21
Lithospheric pressure zone	3
Lithospheric vortex	18
Liquefaction	21
North Banda Basin vortex	8
Pacific Ocean vortex	11
Primary fault	3
Rotating celestial body	1
Sipaliwini vortex	13
Spiral density fault	3
Summary	21
Vortical core	1
Vortical rod	2
Vortical torsion	17

ACRONYMS

DRCB	Differentially Rotating Celestial Body
LV2	Lake Victoria Vortex
NB2V	North Banda Basin Vortex
POV	Pacific Ocean Vortex
RCB	Rotating Celestial Body
SDF	Spiral Density Fault
SV	Sipaliwini Vortex
TDC	Top Dead Center

REFERENCES

Google Earth Pro. (2019).

Lee, J. J. (December 7, 2013). The 2011 Japan Tsunami Was Caused By Largest Fault Slip Ever Recorded. *National Geographic*. Retrieved from https://news.nationalgeographic.com/news/the-2011-japan-tsunami-was-caused-by-largest-fault-slip-ever-recorded/.

Oskin, B. (February 23, 2014). Confirmed: Oldest Fragment of Early Earth is 4.4 Billion Years Old. *Live Science*. Retrieved from https://www.livescience.com/43584-earth-oldest-rock-jack-hills-zircon.html.

Tsuji, Takeshi & Ito, Yoshihiro & Kawamura, Kiichiro & Kanamatsu, Toshiya & Kasaya, Takafumi & Kinoshita, Masataka & Matsuoka, Toshi & Scientists, Shipboard. (2019). Seismogenic Faults of the 2011 Great East Japan Earthquake: Insight from Seismic Data and Seafloor Observations. *Proceedings of the International Symposium on Engineering Lessons Learned from the 2011 Great East Japan Earthquake, March 1-4, 2012. Tokyo, Japan*. Retrieved from https://www.researchgate.net/publication/267427017_SEISMOGENIC_FAULTS_OF_THE_2011_GREAT_EAST_JAPAN_EARTHQUAKE_INSIGHT_FROM_SEISMIC_DATA_AND_SEAFLOOR_OBSERVATIONS.

ACKNOWLEDGEMENTS

Marie Gemmell

Sharolyn Gemmell

Tabitha Fly

Jessica Gemmell

Jamie Nelson

Rick Schwartz

www.ingramcontent.com/pod-product-compliance
Lightning Source LLC
Chambersburg PA
CBHW041211180526
45172CB00006B/1240